I0478802

Contents

Particles

Planck constant ------------ 6
Specific charge -------------- 6
Strong force ------------------ 6
Antiparticles ---------------- 6
Neutrinos -------------------- 6
The electron volt ----------- 7
Feynman diagrams ---------- 7
Similarities ------------------ 7
Standard model ------------- 8
Interactions ----------------- 9

Quantum

Electromagnetic waves ---- 10
Electromagnetic spectrum 10
Photons --------------------- 10
Emissions – gases --------- 10
Florescent tube ----------- 11
Photo-electric effect ------- 11
Stopping potential -------- 11
Wave-particle duality ----- 11
De Broglie wavelength ---- 11

Waves

Definitions ---------------- 12
Transverse/longitudinal -- 12
Resonant frequency ------- 12
Harmonics ----------------- 12
Progressive/standing ----- 13
Polarisation --------------- 13
Diffraction ---------------- 14
Interference --------------- 14
Single slit ----------------- 14
Double slit ---------------- 14
Diffraction grating -------- 14
Snell's law ---------------- 15
Total internal reflection --- 15

Mechanics

Scalars and vectors ------- 16
Components of forces ----- 16
Moments of forces -------- 16
Acceleration --------------- 16
Distance/displacement --- 16
Projectile motion ---------- 17
Suvat ----------------------- 17
Newton's laws of motion -- 17
Work, energy, and power -- 18
Efficiency ------------------ 18
Falling objects ------------ 18

Materials

Solids ---------------------- 19
Density --------------------- 19
Elasticity ------------------- 19
Young modulus ------------ 19

Electricity

Basic equations ----------- 20
Diodes --------------------- 20
Energy and power --------- 20
Special resistors ---------- 20
Circuit rules --------------- 20
Internal resistance -------- 20
Resistivity ----------------- 21
Superconductivity -------- 21

Practical

Planning a practical ------- 22
Hypothesis ---------------- 22
Safety --------------------- 22

[by Walter Brown]

 **Data Sheet**

DATA - FUNDAMENTAL CONSTANTS AND VALUES

Quantity	Symbol	Value	Units
speed of light in vacuo	c	3.00×10^8	m s^{-1}
permeability of free space	μ_0	$4\pi \times 10^{-7}$	H m^{-1}
permittivity of free space	ε_0	8.85×10^{-12}	F m^{-1}
magnitude of the charge of electron	e	1.60×10^{-19}	C
the Planck constant	h	6.63×10^{-34}	J s
gravitational constant	G	6.67×10^{-11}	N m^2 kg^{-2}
the Avogadro constant	N_A	6.02×10^{23}	mol^{-1}
molar gas constant	R	8.31	J K^{-1} mol^{-1}
the Boltzmann constant	k	1.38×10^{-23}	J K^{-1}
the Stefan constant	σ	5.67×10^{-8}	W m^{-2} K^{-4}
the Wien constant	α	2.90×10^{-3}	m K
electron rest mass (equivalent to 5.5×10^{-4} u)	m_e	9.11×10^{-31}	kg
electron charge/mass ratio	$\dfrac{e}{m_e}$	1.76×10^{11}	C kg^{-1}
proton rest mass (equivalent to 1.00728 u)	m_p	$1.67(3) \times 10^{-27}$	kg
proton charge/mass ratio	$\dfrac{e}{m_p}$	9.58×10^7	C kg^{-1}
neutron rest mass (equivalent to 1.00867 u)	m_n	$1.67(5) \times 10^{-27}$	kg
gravitational field strength	g	9.81	N kg^{-1}
acceleration due to gravity	g	9.81	m s^{-2}
atomic mass unit (1u is equivalent to 931.5 MeV)	u	1.661×10^{-27}	kg

ALGEBRAIC EQUATION

quadratic equation $\quad x = \dfrac{-b \pm \sqrt{b^2 - 4ac}}{2a}$

ASTRONOMICAL DATA

Body	Mass/kg	Mean radius/m
Sun	1.99×10^{30}	6.96×10^8
Earth	5.97×10^{24}	6.37×10^6

GEOMETRICAL EQUATIONS

arc length	$= r\theta$
circumference of circle	$= 2\pi r$
area of circle	$= \pi r^2$
curved surface area of cylinder	$= 2\pi rh$
area of sphere	$= 4\pi r^2$
volume of sphere	$= \dfrac{4}{3}\pi r^3$

Particle Physics

Class	Name	Symbol	Rest energy/MeV
photon	photon	γ	0
lepton	neutrino	v_e	0
		v_μ	0
	electron	$e^\pm$	0.510999
	muon	$\mu^\pm$	105.659
mesons	π meson	$\pi^\pm$	139.576
		π^0	134.972
	K meson	$K^\pm$	493.821
		K^0	497.762
baryons	proton	p	938.257
	neutron	n	939.551

Properties of quarks

antiquarks have opposite signs

Type	Charge	Baryon number	Strangeness
u	$+\frac{2}{3}e$	$+\frac{1}{3}$	0
d	$-\frac{1}{3}e$	$+\frac{1}{3}$	0
s	$-\frac{1}{3}e$	$+\frac{1}{3}$	-1

Properties of Leptons

		Lepton number
Particles:	$e^-, v_e \,;\, \mu^-, v_\mu$	$+1$
Antiparticles:	$e^+, \overline{v_e}, \mu^+, \overline{v_\mu}$	-1

Photons and energy levels

photon energy	$E = hf = hc / \lambda$
photoelectricity	$hf = \phi + E_{k\,(max)}$
energy levels	$hf = E_1 - E_2$
de Broglie wavelength	$\lambda = \dfrac{h}{p} = \dfrac{h}{mv}$

Waves

wave speed $\quad c = f\lambda \quad$ period $\quad f = \dfrac{1}{T}$

first harmonic $\quad f = \dfrac{1}{2l}\sqrt{\dfrac{T}{\mu}}$

fringe spacing $\quad w = \dfrac{\lambda D}{s} \quad$ diffraction grating $\quad d \sin \theta = n\lambda$

refractive index of a substance s, $\quad n = \dfrac{c}{c_s}$

for two different substances of refractive indices n_1 and n_2,

law of refraction $\quad n_1 \sin \theta_1 = n_2 \sin \theta_2$

critical angle $\quad \sin \theta_c = \dfrac{n_2}{n_1} \text{for } n_1 > n_2$

Mechanics

moments $\quad$ moment $= Fd$

velocity and acceleration $\quad v = \dfrac{\Delta s}{\Delta t} \quad a = \dfrac{\Delta v}{\Delta t}$

equations of motion $\quad v = u + at \quad s = \left(\dfrac{u+v}{2}\right) t$

$\quad v^2 = u^2 + 2as \quad s = ut + \dfrac{at^2}{2}$

force $\quad F = ma$

force $\quad F = \dfrac{\Delta(mv)}{\Delta t}$

impulse $\quad F\,\Delta t = \Delta(mv)$

work, energy and power $\quad W = F\,s \cos \theta$

$\quad E_k = \dfrac{1}{2} m v^2 \quad \Delta E_p = mg\Delta h$

$\quad P = \dfrac{\Delta W}{\Delta t}, P = Fv$

$\quad efficiency = \dfrac{useful\ output\ power}{input\ power}$

Materials

density $\quad \rho = \dfrac{m}{V} \quad$ Hooke's law $\quad F = k\,\Delta L$

Young modulus $= \dfrac{tensile\ stress}{tensile\ strain}$

tensile stress $= \dfrac{F}{A}$

tensile strain $= \dfrac{\Delta L}{L}$

energy stored $\quad E = \dfrac{1}{2}F\Delta L$

Electricity

current and pd	$I = \dfrac{\Delta Q}{\Delta t}$ $\quad V = \dfrac{W}{Q}$ $\quad R = \dfrac{V}{I}$
resistivity	$\rho = \dfrac{RA}{L}$
resistors in series	$R_T = R_1 + R_2 + R_3 + \ldots$
resistors in parallel	$\dfrac{1}{R_T} = \dfrac{1}{R_1} + \dfrac{1}{R_2} + \dfrac{1}{R_3} + \cdots$
power	$P = VI = I^2 R = \dfrac{V^2}{R}$
emf	$\varepsilon = \dfrac{E}{Q} \qquad \varepsilon = I(R + r)$

Key

equation

Definition

Extracts from mark scheme

Particles

Planck constant

$$e = hf$$ $$h = 6.63 \times 10^{-34}$$

Specific charge

$$specific\ charge = \frac{charge}{mass}$$

Unit: Ckg^{-1}

Strong force

Three important facts
1. Does not depend on charge:
   It's the same between p-p, n-n, and p-n
2. Only occurs when particles are very close together
   Only occurs in the nucleus for this reason
3. If nucleons become too close, they repel
   Otherwise the nucleus would collapse

Attraction/repulsion
Attraction: $0.0 - 0.5\ fm$ Repulsion: $0.5 - 3.0\ fm$

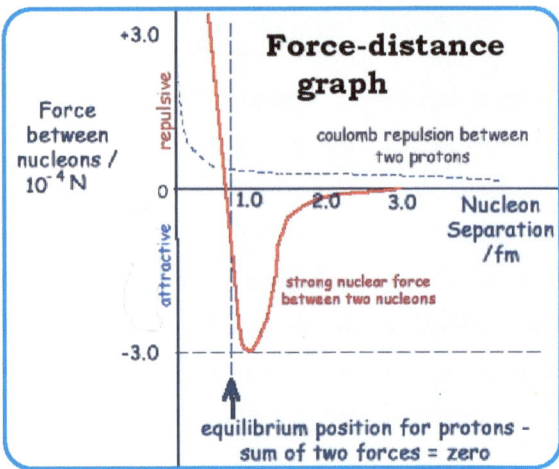

Force-distance graph

Force between nucleons / 10^{-4} N

repulsive / attractive

coulomb repulsion between two protons

strong nuclear force between two nucleons

equilibrium position for protons - sum of two forces = zero

Antiparticles

For every particle there is a corresponding antiparticle that:
- Has the opposite charge
  (if the particle has a charge)
- Has exactly the same rest mass as the particle
- Has the opposite quark structure
- Will annihilate the particle and itself if they meet, converting their total mass into photons.

Neutrinos

When beta decay was first discovered, it appeared that the beta particles could have a range of energies, even when coming from identical nuclei. Either the rule of conservation of energy had been broken or a mysterious particle was carrying the missing energy. The particle was called the neutrino. They have no charge, hardly any mass, and billions of them from the sun pass through us every second.

Rest energy
Energy is equivalent to mass, shown by $E = mc^2$. The mass of a particle increases as it moves faster. The energy equivalent of it's mass at rest is called *rest energy*.

PET scan
- *Positron Emission Tomography*
- A positron emitting isotope is put into the blood stream and some reaches the brain.
- The positron and it's antiparticle the electron annihilate before the positron travels a few mm.
- Gamma photons are produced and a detector is used to build up a picture of the brain.

Annihilation
Occurs when a particle and it's antiparticle meet and their mass is converted into radiation energy.
Two photons produced (to conserve momentum).
Minimum energy of each photon produced (E_0 is the rest energy of each particles):

$$hf_{min} = E_0$$

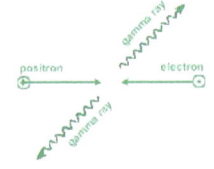

Pair production
A photon with specific energy can suddenly change into a particle antiparticle pair.
Minimum energy a photon must have to create a particle-antiparticle pair each of rest energy E_0 is:

$$hf_{min} = 2E_0$$

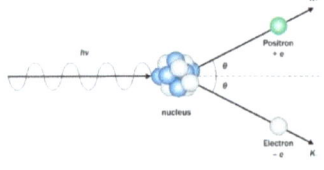

The electron volt

$$1eV = 1.6 \times 10^{-19}J$$

An **electron volt** is the energy transferred when an electron is moved through a potential difference of 1 volt

Similarities

Similarities between all particles
- *All have rest mass*
- *All affected by weak interaction*
- *If charged all experience electromagnetic interaction*

Feynman diagrams

These mean nothing:
Lengths of lines, angles, sizes.
These are conserved at vertices:
Charge, energy, momentum, baryon number, lepton number, *(sometimes strangeness)*

Strangeness
Strangeness is **conserved** in interactions where strange particles are **made**.
Strangeness is **not conserved** in interaction where strange particles **decay**.

Standard model

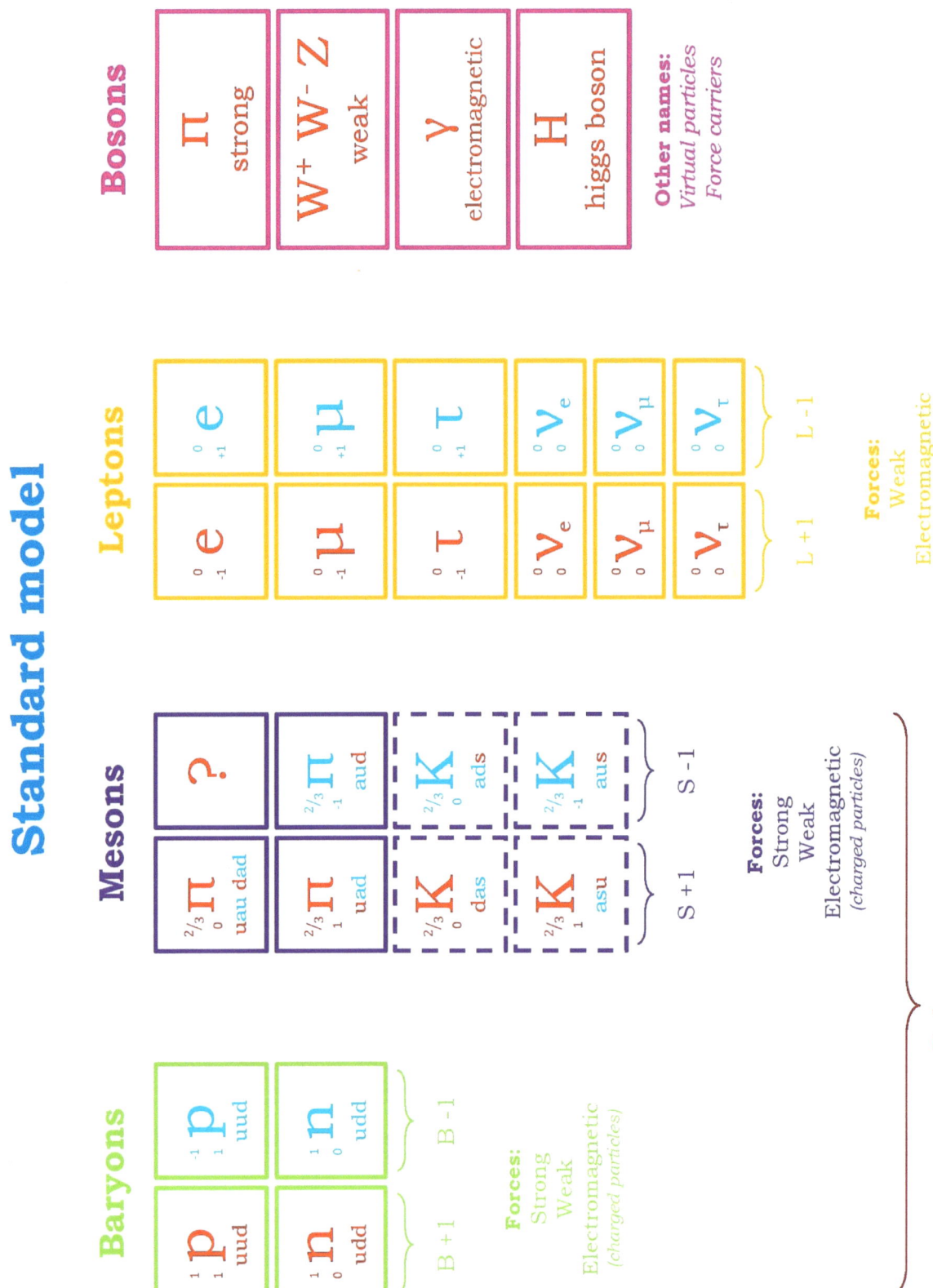

Interactions

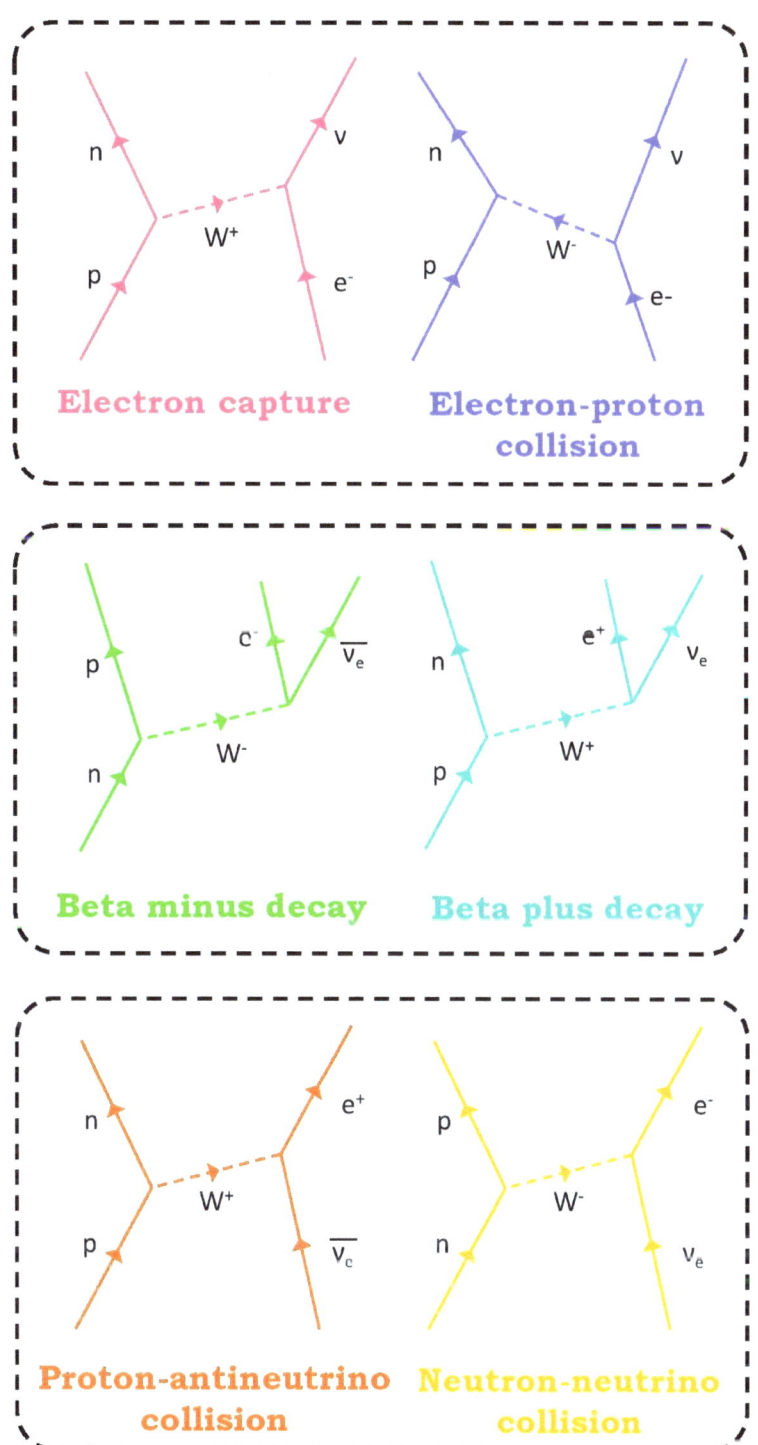

Electron capture

Electron-proton collision

Beta minus decay

Beta plus decay

Proton-antineutrino collision

Neutron-neutrino collision

Quantum

Electromagnetic waves

Electromagnetic waves all:
- Caused by oscillating charge
- Transverse waves
- Travel at 3×10^8 ms^{-1} in a vacuum

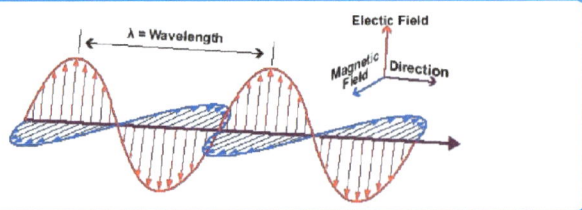

Electromagnetic spectrum

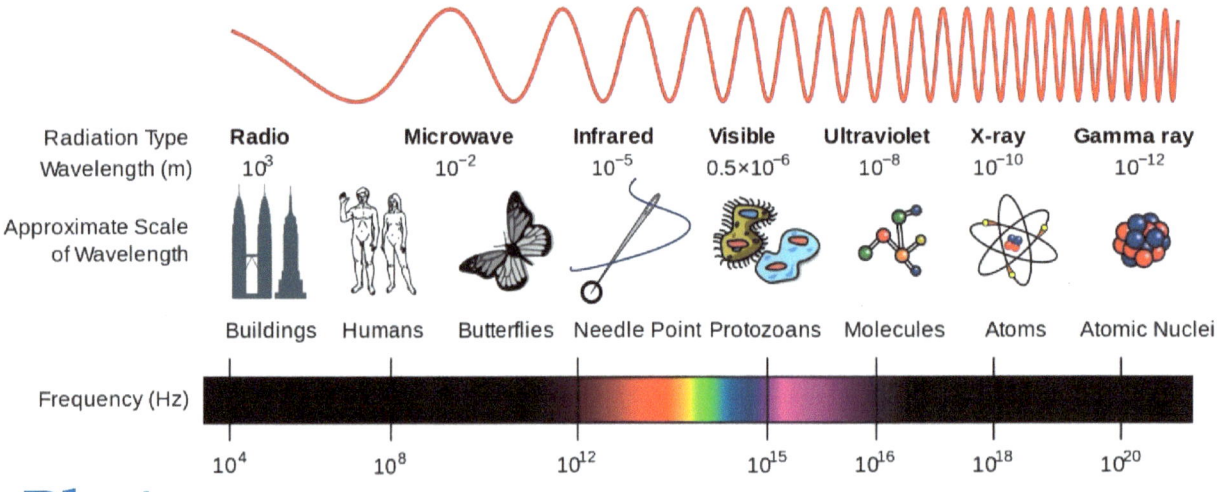

Radiation Type	Radio	Microwave	Infrared	Visible	Ultraviolet	X-ray	Gamma ray
Wavelength (m)	10^3	10^{-2}	10^{-5}	0.5×10^{-6}	10^{-8}	10^{-10}	10^{-12}
Approximate Scale of Wavelength	Buildings	Humans	Butterflies	Needle Point	Protozoans	Molecules	Atoms	Atomic Nuclei

Frequency (Hz): 10^4 10^8 10^{12} 10^{15} 10^{16} 10^{18} 10^{20}

Photons

Light travels in wave packets called photons.
The energy of a photon depends on the frequency of the light or radiation.

$$E = hf$$

Emissions - gases

Collisions of electrons and atoms
1. **Elastic** – KE conserved during collision
2. **Inelastic** – Some KE absorbed by atom
   a. **Excited** – gives electron enough energy to move up an energy level. When it drops back down, a photon is released.

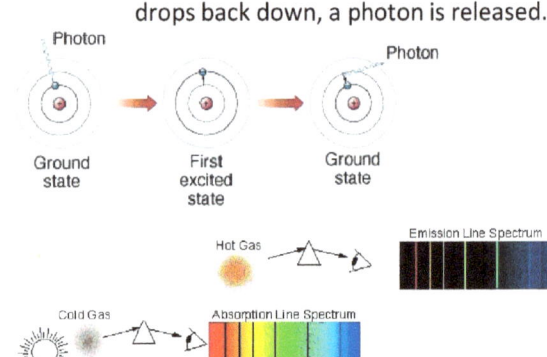

   b. **Ionised** – gives electron enough energy to escape from atom.

Reasons for emission spectra
Collisions
1. *Energy from collision of charged particles transfers to electrons in gas molecules*
2. *Electrons excited to higher energy levels*
3. *The more energy the electrons absorb the higher the energy levels reached*
4. *Electrons are unstable at higher energy levels so will fall back down, emitting a photon*

Formation of spectral lines
1. *Photon energy = hf (energy $\propto$ frequency)*
2. *Spectral lines are at specific wavelengths – atoms have discrete (quantised) energy levels*
3. *Each spectral line corresponds to an electron falling down to a lower energy state*
4. *Energy gap, $\Delta E = hc/\lambda$*
5. *Larger energy gap means higher energy photon is emitted so shorter wavelength or vice versa*
6. *Many frequencies because electrons return to lower levels by different route*
7. *Photons of characteristic frequencies emitted from atoms of a particular element*

Florescent tube

How it works:
- The tube contains mercury vapour at low pressure.
- Electrons are fired through the tube (negative end, positive end)
- Fast moving electron causes an electron in a mercury atom to rise up an energy level (just the right amount of energy)
- The excited electron drops back down to its original energy level, emitting a UV photon.
- This UV photon hits electron in phosphorous atom in coating.
- Electron drops down again and emits a proton. (Yellow frequency)

Photo-electric effect

Put simply:
Light causes electrons to be ejected from the surface of a metal.
$E = hf$ so photons with a higher frequency have higher energy

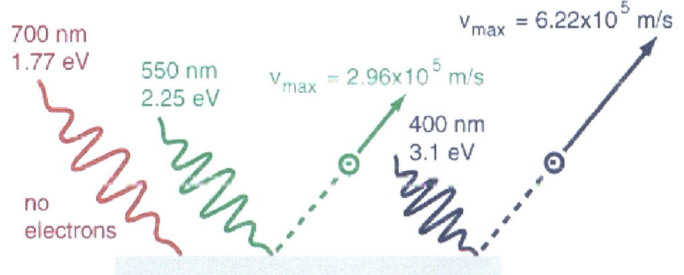

700 nm
1.77 eV
550 nm
2.25 eV
$v_{max} = 2.96 \times 10^5$ m/s
$v_{max} = 6.22 \times 10^5$ m/s
400 nm
3.1 eV
no electrons

Potassium - 2.0 eV needed to eject electron

Work function
Threshold frequency (f_0) is the minimum frequency to cause photoemission.
No electrons emitted below threshold frequency because:
- *Energy of a photon depends on frequency ($E = hf$)*
- *Below threshold frequency photon does not have enough energy to liberate an electron*

Work function = $hf_0 = \Phi$ = minimum energy to cause photoemission

$$hf = hf_0 + KEmax$$

Stopping potential

Put simply:
- The emitter gives out electrons. So it's called a cathode.
- It causes a current to flow in the circuit.
- But as we increase the e.m.f. of the power supply, the emitter becomes positively charged (it's connected to the positive terminal of the supply)
- So electrons leaving it are attracted back towards it.
- By increasing the e.m.f. of the supply you can find the p.d. where no electrons are able to cross the gap.
- Even those with maximum energy, E_{kmax}, can't do it.

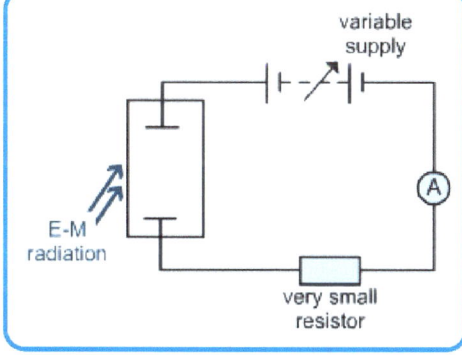

variable supply
E-M radiation
very small resistor

$$eV_s = Ekm_{ax} = \frac{1}{2} mv^2$$

Where: e = charge on electron V_s = stopping potential
m = mass of electron v = speed of electron

Wave-particle duality

	Electrons	Light
	Carry charge Bend in magnetic field Maltese cross	Photoelectric effect Sharp shadows Photons
	Electron ring pattern in discharge tube – diffraction	Diffraction No mass, travels at "c" Double slit experiment

De Broglie wavelength

$$\lambda = \frac{h}{mv}$$

This equation can be used to calculate the wavelength of any particle.
Where h is the Planck's constant.

Waves

Definitions

Wavelength λ (m) is the distance between two adjacent crests

Amplitude A (m) is the maximum displacement from the mean position. Waves with a bigger amplitude carry more energy.

Velocity v (ms^{-1}) is the distance moved by a given point on the wave in 1 second

Frequency f (Hz) is the number of complete wavelengths passing a point in a second

$$f = \frac{1}{T}$$

Period T (s) is the time for one wavelength to pass

$$v = f\lambda$$

Transverse / longitudinal

In **Transverse waves** the direction of travel is perpendicular to the direction of oscillation.

Examples:
- Light
- Rope / Slinky
- Surface of water

[Can be polarised] [Does not need medium]

In **Longitudinal waves** the direction of travel is parallel to the direction of oscillation.

Examples:
- Sound waves
- Seismic waves
- Deep water

[Cannot be polarised] [Needs medium]

Resonant frequency

$$f = \frac{\sqrt{T}}{2l\sqrt{\mu}}$$

f = frequency (Hz)
T = tension (N)
l = length of vibrating spring

Harmonics

Harmonic	Frequency	Open pipes / strings			Closed pipes		
		Node positions	λ	Overtone	Node positions	λ	Overtone
1	f_1		$2l$	Fund. Freq.		$4l$	Fund. Freq.
2	$2f_1$		l	1st			
3	$3f_1$		$\frac{2}{3}l$	2nd		$\frac{4}{3}l$	1st
4	$4f_1$		$\frac{1}{2}l$	3rd			
5	$5f_1$		$\frac{2}{5}l$	4th		$\frac{4}{5}l$	2nd

Progressive / standing waves

Progressive waves
- All points in the wave have equal amplitude
- All the points along the wave have different phase
- Points one wavelength apart are "in phase" with each other
- They are only produced in unbounded mediums (mediums without boundaries.)
- They transfer energy but the medium is not transferred.

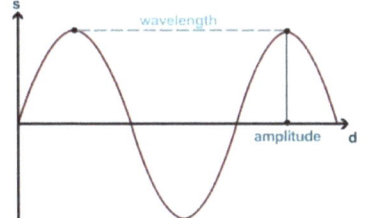

Standing waves
- The amplitude varies from zero at nodes to maximum at antinodes
- All particles between nodes are in phase
- The wave does not move through the medium
- No energy is transferred along the wave

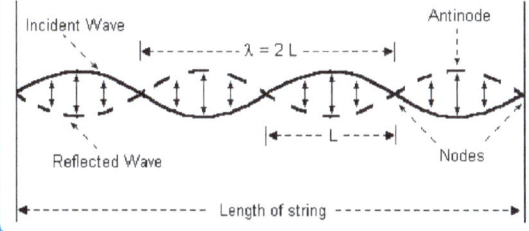

Phase difference

Phase difference is the fraction of a cycle between the vibrations of two different waves, in degrees or radians

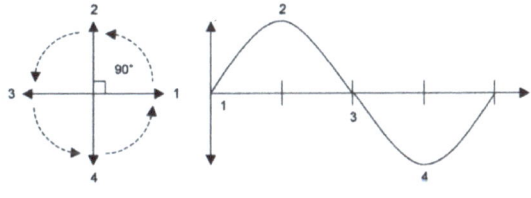

Formation of standing waves
- Waves reflect off the clamp (and the rod)
- Wave meets a reflected wave
- Same wavelength or frequency, similar amplitude
- Superposition of the two waves
- Node
  - point of minimum or no disturbance
  - cancellation / destructive interference
  - 180° phase difference
- Antinode
  - point of maximum disturbance
  - reinforcement / constructive interference
  - in phase
- Between node and antinode, amplitude of oscillation increases
- Energy is not transferred along string

Polarisation

Polarised light

Polarised light oscillates in 1 plane only

Only transverse waves can be polarized

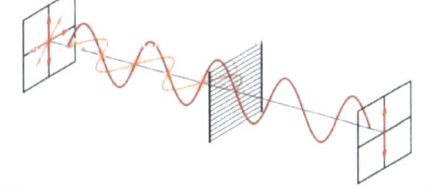

Polarising filter
Polarisation filters are specially adapted glass that, when turned at an angle to a light source, will reduce glare from reflected surfaces.
Uses:
- Sunglasses / windscreens to reduce glare
- Camera to enhance image
- Microscope to identify minerals/rocks

It is important to align a TV aerial correctly, as transmitted waves are often polarised, and aerial rods must be aligned in the same plane of polarisation of the wave.

Diffraction

Diffraction is the bending of a wave when it passes through an aperture or round an obstacle that is similar in size to it's wavelength

Interference

Principle of superposition
The overall displacement caused when two waves meet is equal to the sum of the displacements from each individual wave

Coherent sources
- Same frequency and wavelength
- Have a constant or zero phase difference
- Have roughly the same amplitude
- Polarised in the same plane

Single slit

$$d \sin \theta = n\lambda$$

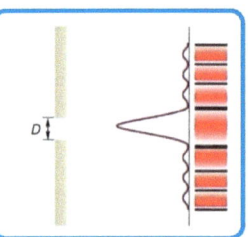

To find **Minima**
d=slit width
θ=angle to minima

Notes
- The overall displacement caused when two waves meet is equal to the sum of the displacements from each individual wave
- Narrower slit = wider pattern + higher intensity

Will diffraction occur?
- *How many times the wavelength is the slit width*
- $\sin \theta = \lambda/d$
- *If angle greater than 5° diffraction will occur*

Double slit

$$wS = \lambda D$$

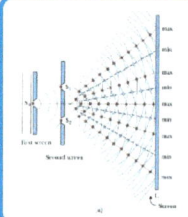

To find **Maxima**
D=distance between maxima
S=distance between Slits

Problems caused
- Less detail visible to human eye. Solution: use UV light
- Ghosting on T.V.

White light	Red Laser
Different colours and central white fringe	*One colour*
Less intense	*More intense*
Maxima wider	*Maxima narrower*
Max/min closer together	*Max/min further apart*
Fringes	*Dots*

Diffraction grating

$$d \sin \theta = n\lambda$$

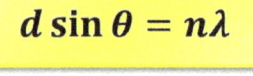

To find **Maxima**
d=distance between slits
θ=angle to maxima

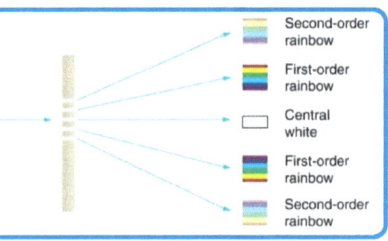

$$n = \frac{d}{\lambda}$$

n = total number of orders

Used to: evaluate elements from distant stars

Drawing fringes of white light
- *Distinct fringes with subsidiary maxima*
- *Colours are present within each subsidiary maxima*
- *Blue inner edge and red outer edge*
- *Central white maximum*

Diffraction grating vs double slit
When calculating wavelength, it is more sensible to use a diffraction grating rather than a double slit, as:
- *A brighter better-defined pattern is produced*
- *Spectra are further apart so angle can be measured more accurately.*

Snell's law

$$n_1 \sin \theta_1 = n_2 \sin \theta_2$$

$$_1n_2 = \frac{c_1}{c_2} = \frac{\lambda_1}{\lambda_2} = \frac{n_2}{n_1}$$

In air:

$$n_2 = \frac{\sin i}{\sin r}$$

When the refractive indexes are closer together, less bending occurs.

$_1n_2$ is beam travelling from 1 to 2

Total internal reflection

Critical angle

Total internal reflection occurs at the boundary between two media when:

- *The light ray is in the optically denser medium, travelling into a medium that is less optically dense*
- *When the angle of incidence is bigger than the critical angle*

$$\sin C = \frac{n_2}{n_1}$$

It is an advantage to have a small critical angle

Optical fibres

In a step-index multi-mode fibre, rays of light are guided along the fibre core by total internal reflection. The critical angle is determined by the difference in refractive index between the core and cladding materials.

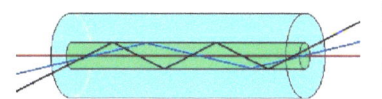

Step index

Monomode

Advantage of smaller diameter core

- *Less multipath dispersion*
- *Better quality signal / less smearing*
- *Increased probability of TIR*
- *Less change of angle → angle of incidence less likely to fall below the critical angle → less refraction out of the core*

Reason for cladding

- *Protects the core from scratches*
- *Increases the critical angle*
- *Reduces pulse broadening*
- *Reduces modal dispersion*
- *Increases rate of data transfer*

Signal decay

- *Reduced amplitude*
  - *Absorption*
  - *Energy loss in fibre*
  - *Attenuation (reduction of signal strength)*
- *Pulse broadening*
  - *Multi-path dispersion*
  - *Different rays*
  - *Modes propagating at different angles —greater angles take more time*
  - *Smearing of signal*

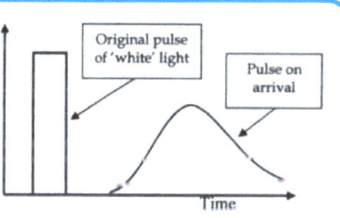

Original pulse of 'white' light

Pulse on arrival

Time

Mechanics

Scalars and vectors

A **scalar** has only magnitude	A **vector** has magnitude and direction

Scalars – examples
Distance, Speed, Energy, Power, Voltage, Temperature

Vectors – examples
Displacement, Velocity, Acceleration, Force, Weight, Momentum

Adding vectors
Method 1: Lay the vectors nose to tail
Method 2: Parallelogram
Method 3: Using trig (only if vectors are at right angles)
If vectors form a closed loop, they are perfectly balanced.

Direction matters

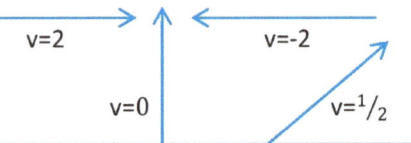

Components of forces

$P \cos a + Q \cos b = mg$	$F \cos \theta = P$	$F \sin \theta = Q$

Finding component – no right angle

P   Q
a   b
W=mg

Finding component – with right angle

Remember:
Cos is "cosy" with the line.

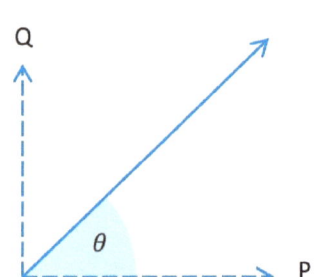

Q θ P

Moments of forces

For a body in equilibrium, the total clockwise moment about any point is equal and opposite to the total anticlockwise moment about the same point.

Moment (Nm) = F × s

Couples
Moment of a couple = Fs

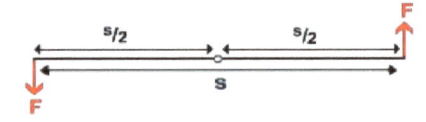

Acceleration

$A = \dfrac{\Delta v}{t}$

Acceleration-time graphs
Straight line:
a (in/de)creasing
Flat: constant a
Negative: deceleration

Distance/ Displacement

s (Displacement)
d (Distance)

Projectile motion

A **Projectile** is an object only acted upon by the force of gravity

Rules
- Always accelerating downwards at $9.81ms^{-2}$
- Their motions vertically and horizontally are completely independent (do not affect each other)
- They have a constant horizontal velocity
- t is the only link between their horizontal and vertical motion.

Solving horizontally
Use $s = vt$
(no acceleration)
Solving vertically
Use **SUVAT**
(acceleration)

Suvat

$$v = u + at$$ s missing

$$s = ut + \frac{1}{2}at^2$$ v missing

$$s = \frac{1}{2}(v + u)t$$ a missing

$$v^2 = u^2 + 2as$$ t missing

Newton's laws of motion

Newton's 1st law

Every object will continue in uniform velocity or stationary unless acted on by an external resultant force

Resultant forces can cause objects to change velocity (including direction) or change shape.

Newton's 2nd law

Force is equal to the rate of change of momentum

$$F = \frac{mv - mu}{\Delta t}$$

$$F = ma$$

If a resultant external force acts on an object, it will accelerate. More massive objects require larger forces to accelerate.
Force time graphs:

Area under graph = impulse = Δmomentum

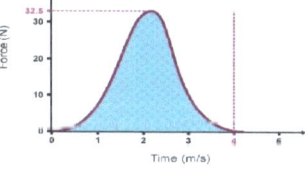

Within a closed system, with no external forces acting on it, momentum is conserved

Newton's 3rd law

Every action has an equal and opposite reaction

What the object feels is the same as what the object pushing it feels, even if there is a resultant force.

Work, energy, and power

$$p = \frac{E}{t}$$

Power is the rate of energy transfer

$W = Js^{-1}$

$$p = Fv$$

Work
- Work is done when energy is transferred
- If something can supply or feel a force, it can do work

A **field** is a region around an object that will cause another object to feel a force

$$W = Fs$$

Work done at an angle

$$W = Fs\cos\theta$$

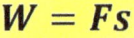

F is a force
s is the direction of travel

Efficiency

Key points
No machine or energy transfer is 100% efficient
Energy lost:
- *Energy is transferred to: heat/sound*
- *Friction (be specific)*
- *Drag / air resistance*
- *Work done against or by a resistive force*

E_{GP} converted to E_K
Happens when an object moves in a gravitational field

$$v^2 = u^2 + 2as$$
$$v^2 = 2gh$$
$$v = \sqrt{2gh}$$

Falling objects

Velocity is independent of mass
Falling object, negligible air resistance
- *All GPE lost is transferred to KE (no GPE transferred to heat)*
- $mg\Delta h = \frac{1}{2}mv2$
- *mass cancels*
- $v = \sqrt{2gh}$

Drag and terminal velocity
- *Acceleration is maximum (9.81ms-2) at the beginning*
- *Velocity increases and then becomes constant (terminal velocity)*
- *Two forces are weight and drag*
- *Weight initially greater than drag*
- *Resultant force ($W - drag$) causes acceleration*
- *Drag force increases with speed*
- *Forces become balanced*
- *Acceleration reduces to zero*

Materials

Solids

Particles
In a solid the particles (either atoms or molecules) vibrate about a fixed position. The amplitude of the vibration depends upon their temperature. The links or bonds between particles are permanent and there is normally long range order within the structure.

Types of solids
- **Crystalline**
    - Ordered patterns
    - Simple structure called a unit cell is repeated in a 3D pattern.
    - E.g. metals, crystals
- **Amorphous**
    - Only short range order
    - Formed from supercooled liquids (i.e. fast cooling)
    - E.g. glass, some plastics
- **Polymeric**
    - Consists of long chains of a simple structure repeated over and over again
    - E.g. many organic materials, plastics

Density

What is density?
Density is a way of describing how much mass a material contains in a given volume
Units: $Kg\,m^{-3}$

$$\rho = \frac{m}{v}$$

Deriving $E = \frac{1}{2}F\Delta L$ from F/l graph
- *($\Delta W = F\Delta s$) so area beneath line from origin to ΔL represents work*
- *work done linked to energy stored*
- *area of triangle = ½ $b \times h$ therefore $E = $ ½ $F\Delta L$*

Elasticity

Types of force
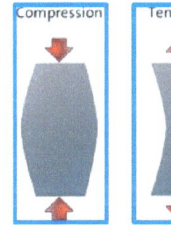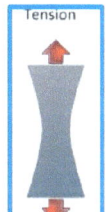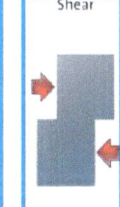
Compression | Tension | Shear

$$\frac{F}{\Delta L} = k \left(\begin{array}{c} spring \\ constant \end{array} \right)$$

Hysteresis
Some materials do not give back all the energy they are given when stretched.

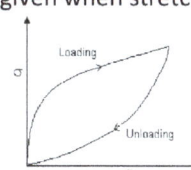

This extra energy is absorbed by the material as heat.

Energy stored in a stretched material
When an object is stretched by a force, work is done. As a result, energy is transferred to and stored in the material.

$$elastic\ energy = {}^{1}\!/_{2}\ F\Delta l$$

But $F = k\Delta l$, so:

$$elastic\ energy = {}^{1}\!/_{2}\ kl^{2}$$

Hooke's law
When a material is put under tension, the amount it stretches (extension) is proportional to the force applied (load).

Stress-strain graph
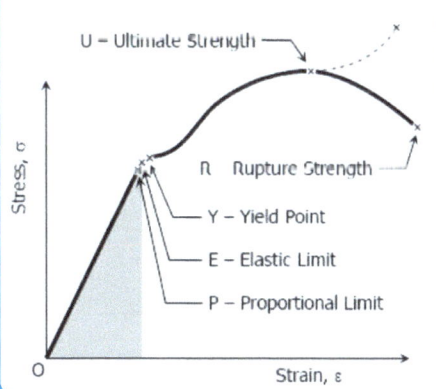

- U – Ultimate Strength
- R – Rupture Strength
- Y – Yield Point
- E – Elastic Limit
- P – Proportional Limit

Up to **P**, the material obeys Hooke's Law

After **E**, the material does not return to its original shape

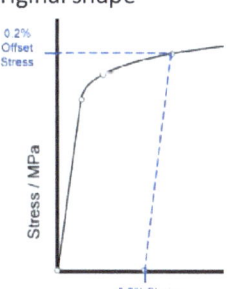
0.2% Offset Stress
0.2% Strain

Young Modulus

$$E = \frac{Young}{modulus} = \frac{stress}{strain} = \frac{\sigma}{\varepsilon} = \frac{Fl}{eA}$$

Units: Nm^{-2}

$$stress = \frac{F}{A}$$
Units: Nm^{-2}

$$strain = \frac{\Delta l}{l}$$
Units: none

Typical values ($\times 10^{10}$)
Rubber: 0.005; Glass: 7; Steel: 21

Electricity

Basic equations

$$V = I \times R$$

$$E = P \times t$$

$$P = V \times I$$

$$E = Q \times V$$

$$Q = I \times t$$

Diodes

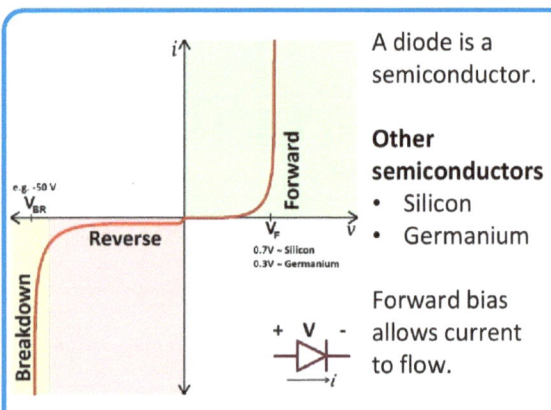

A diode is a semiconductor.

Other semiconductors
- Silicon
- Germanium

Forward bias allows current to flow.

Energy and power

$$P = I^2 R$$

$$P = \frac{V^2}{R}$$

$$E = VIt$$

A Watt is a Joule per second

Special resistors

Light dependent resistors
Light → Lower resistance
Thermistors
Heat → Lower resistance

Circuit rules

Series circuits
- The current is the same at all points in the circuit
- The p.d. is shared between components
- Larger resistance takes more p.d.
- The total voltage of cells in series is the sum of their p.d.s.
- The total resistance of resistors in series is the sum of their individual resistances.

Parallel circuits
- The total current entering or leaving the supply is the sum of the currents in each branch
- The total p.d. across each branch is the same
- Branches with bigger resistance take less current

$$\frac{1}{R_T} = \frac{1}{R_1} + \frac{1}{R_2} + \frac{1}{R_3}$$

- For identical resistors:

$$R_T = \frac{R_1}{n}$$

Internal resistance

Key points
- This is the resistance of the battery
- We can't measure r

Emf is the potential difference across the terminals of the cell when no current is flowing

- We can find r, by plotting V against I, varying I with a variable resistors.

$$E = V + Ir$$

Graph

$$V = -rI + E$$

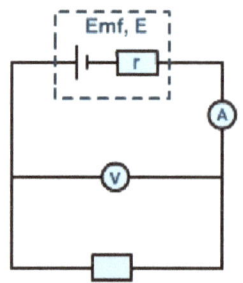

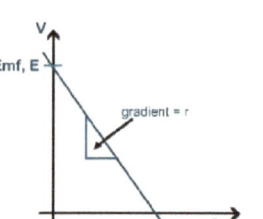

Resistivity

A constant for a material

The electrical resistance of a metal wire depends on:

- It's length, $l\ (m)$
- It's cross sectional area, $A\ (m^2)$
- It's resistivity, $\rho\ (\Omega m)$

$$R = \frac{\rho l}{A}$$

Typical resistivity values

Type	Material	Resistivity (Ωm) @ 20°C
Conductors	Silver	1.6×10^{-8}
	Copper	1.7×10^{-8}
	Aluminium	2.8×10^{-8}
	Constantan	4.9×10^{-7}
Semi-conductors	Germanium	4.2×10^{-1}
	Silicon	2.6×10^{3}
Insulators	Polythene	2×10^{11}
	Glass	$\sim 10^{12}$
	Epoxy resin	$\sim 10^{15}$

Superconductivity

Intro

When certain specialised materials are cooled below a critical temperature, T_c , they lose all resistance.

Uses

A current flowing through such a material causes no heating effect on it and so there is no voltage drop across the material. Once set in motion a current in a superconductor will flow forever. Normally electromagnets waste a lot of energy as heat because of the very large currents needed. Super-conducting electromagnets can have huge currents making very strong magnetic fields without needing high voltages to maintain them or producing any waste heat. Used in MRI machine, or MAGLEV levitating train.

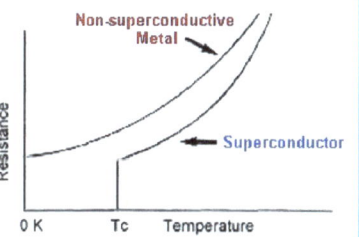

Practical

Planning a practical in the exam

- Correct equipment set up
- Correct circuit diagram
- Specific measurement – use right equipment
- Vary .... + take reading ....
- Give range of measurement
- Repeat – 3 sets of results (mean)
- Draw a graph of .... vs ....
- My graph will show (gradient, y intercept)
- Accuracy
- Specific issue: precision

Hypothesis

- Hypothesis needs to be tested by experiment
- Experiment must be repeatable
- Hypothesis is then accepted or rejected

Safety

Lasers
- Don't shine towards a person
- Avoid accidental reflections
- Wear laser safety goggles
- 'Laser on' warning light outside room
- Stand behind laser